W9-AXO-118

Musical instruments

Dorothy Diamond and Robert Tiffin

A Chelsea College Project sponsored by the Nuffield Foundation and the Social Science Research Council

Published for Chelsea College, University of London, by Macdonald Educational, London and Milwaukee

First published in Great Britain 1976 by
Macdonald Educational Ltd
Holywell House, Worship Street
London EC2A 2EN

Macdonald-Raintree Inc
205 W. Highland Avenue
Milwaukee, Wisconsin 53203

Reprinted 1977, 1978, 1980, 1981

ISBN 0 356 05077 7

Library of Congress Catalog Card Number
77-82979

Project team

Project organizer: John Bird

Team members: Dorothy Diamond
                Keith Geary
                Don Plimmer
                Ed Catherall

Evaluators: Ted Johnston
                Tom Robertson

Editor

Penny Butler
Macdonald Educational

with the assistance of
Nuffield Foundation Science Teaching Project
Publications Department

Filmset by Layton-Sun Ltd

Made and printed by
Morrison & Gibb Ltd, London and Edinburgh

General preface

The books published under the series title Teaching Primary Science are the work of the College Curriculum Science Studies project. This project is sponsored jointly by the Nuffield Foundation and the Social Science Research Council. It aims to provide support and guidance to students who are about to teach science in primary schools.

Although the College Curriculum Science Studies materials have been produced with the student teacher very much in mind, we suggest that they will also be of use to teachers and to lecturers or advisers —in fact to anyone with an interest in primary school science. Hence this series of books.

Three main questions are considered important:

What is science?

Why teach science?

How does one teach science?

A very broad view is taken of teacher training. Training does not, and should not, stop once an in-service or college course has been completed, but can and does take place on a self-help basis in the classroom. In each context, however, we consider that it works best through the combined effects of:

1 Science Science activities studied practically at the teacher's level before use in class.

2 Children Observation of children's scientific activities and their responses to particular methods of teaching and class organization.

3 Teachers Consideration of the methods used by colleagues in the classroom.

4 Resources A study of materials useful in the teaching of science.

5 Discussion and thought A critical consideration of the *what*, the *why* and the *how* of science teaching, on the basis of these experiences. This is particularly important because we feel that there is no one way of teaching any more than there is any one totally satisfactory solution to a scientific problem. It is a question of the individual teacher having to make the 'best' choice available to him in a particular situation.

To help with this choice there are, at frequent intervals, special points to consider; these are marked by a coloured tint. We hope that they will stimulate answers to such questions as 'How did this teacher approach a teaching problem? Did it work for him? Would it work for me? What have I done in a situation like that?' In this way the reader can look critically at his own experience and share it by discussion with colleagues.

All our books reflect this five-fold pattern of experiences, although there are differences of emphasis. For example, some lay more stress on particular science topics and others on teaching methods.

In addition, there is a lecturers' guide *Students, teachers and science* which deals specifically with different methods and approaches suitable for the college or in-service course in primary science but, like the other books in the series, it should be of use to students and teachers as well as to lecturers.

Contents

Introduction

Why choose musical instruments?

The material is attractive, and the activities have a value beyond the science they include, as well as being available to all.

The apparatus can play a significant part both in school life and in everyday experience, unlike some of the things used in laboratories which smell of unreality and science fiction. Musical instruments are excellent experimental material, leading easily to discovery physics which can be understood and enjoyed from the first year in the primary school.

Since every school has, or can have, the simple instruments needed, music offers good opportunities for non-science teachers and for those who do not specialize in music to work with children, encouraging the senses of hearing, touch and sight, as well as musical ability and logical thought.

The topic is one which leads into and can branch out of many important areas of education. It can bridge gaps, too, so that children can see patterns instead of fragmented ideas.

Teachers using musical instruments with primary school children may well ask 'Why teach science?' Here are some answers:

To help children to practise good observation: the hearing ear and the seeing eye together.

To encourage them to think, linking cause with effect and checking results, not accepting statements without evidence.

To show them how to test practically; to find out for themselves which change among several variable factors has which result.

To permit them to measure, where measurement is significant and within their ability.

To help them to see worthwhile problems to solve, and to find practical ways of solving them.

These aims are not only part of basic scientific method, but are clearly important activities for the growing intelligence.

1 Beginning

Every primary class is likely to have some experience of what could loosely be called music-making.

Apparatus and organization

What do they have already and what do they need?

Instruments

Infants probably have in their room percussion instruments, perhaps shakers, a triangle and striker and a tambourine to accompany melodies on record or tape. They may have much more, possibly even a kind of glockenspiel (they are unlikely to have a real wooden xylophone) and tubular or hand bells, recorders, a teacher's guitar. They may, too, have access to the school piano—perhaps even one in the classroom.

Juniors will probably have all these things and more.

Materials Besides recognized musical instruments the class will need:

Bamboo garden canes
Coffee jars
Kitchen towel-roll tubes
Tape measures
Rubber bands, plenty, different sizes

Yoghurt pots
Dried peas
Wood (batten, dowel rod)
One or two basic tools
Sellotape
Boxes

> Probably almost all of this is already in the classroom. Go through a few chapters of this book and see what else you will need, if anything.

Storage Where will you put (a) materials, (b) work under construction, (c) instruments owned by the school or privately? A music science trolley is a good solution. Some of the objects will be breakable, some space-demanding, and it will be a great help to be able to wheel the load to the place where it is to be used, and away again. The instruments should be kept in a lockable cupboard or store-room.

Space Where can children work and play instruments at the same time? There may be a small room off the classroom, or an enclosable area in an open plan scheme. The best place might be a corner of the classroom to which the trolley can be wheeled.

Noise What can you and the rest of the school tolerate? How much noise are the children likely to generate? Guided discoveries and scientific testing do not sound like choir practice or the recorder ensemble. This needs thought and perhaps discussion.

Grouping Both the use of the available musical instruments and the amount of sound likely to arise call for planning. Probably the children should be divided into groups for this work.

There are, though, some activities in which a whole class can be suitably and happily involved, such as listening to a tape-recording (see Chapter 10) or discussing vocabulary (Chapter 7). Most of the guided discovery work should be personal experience in a small group, since all the children want, and indeed should want, to have a go.

> How do you group the children for work in which their backgrounds may be so widely different? Do you put able children with less able, all musicians together, groups of friends together?

Example Brian (aged five and a half) was not concerned with these problems. At every chance of movement, for example in number work round his crowded infant classroom, he would pop over to the small glockenspiel at the side and play a few quick notes—no melody, just a bit of 'music' for fun. Sensibly, the student teaching the class took no special notice, so that this did not become a conflict. Everybody in the class heard it, and as the door was often open, so did other people. Suppose six of the class had all had the same idea, or it had turned into a prestige activity?

What would you advise? Take the glockenspiel away? Provide five more? Or what? This one is a general classroom problem. What happens when the attractive object is something more disruptive?

What solutions do you use in this kind of situation? What alternatives could be considered?

Instruments with a fixed pitch, such as the glockenspiel, help the untrained, the unskilful, the unconfident, the uncoordinated and the under-privileged children to get it right. Brian could not have dealt with any instrument where he needed to make the pitch, or identify it.

A sounds investigation table (see page 6)

Workcards

To help individuals Pupils differ very widely in their musical ability and experience even within a narrow age range in one school. For instance, some may have a parent in the operatic society, others an elder brother in a jazz or pop group; a girl might be taking piano lessons or a boy singing in a choir.

Such unevenness makes workcards an enormous help in classroom organization. Those who are able to go fast can do so, and nobody need be bored.

The able teacher might be able to cram everything which needs to be said into a lecture-type lesson. However, for the child who misses a word or two, or is absent, there is no way of going back. Workcards are more permanent and more readily available than the spoken word. This makes for confidence and continuity.

For activities with instruments Musical instruments of all but the most unsophisticated kinds are expensive, and few schools would spend their allowance for years on, say, a class set of xylophones. Hence workcards for activities with various instruments will allow flexibility and permit everything to be used fully. Without them the teacher may be too busy to help and stimulate every child.

For getting the best from instruments
Things, such as musical instruments, which respond to human actions are particularly attractive. Getting the most satisfactory response from a musical instrument may, however, depend on quite a small factor, such as hitting a tuning-fork in the right way and then holding it correctly (see page 35). Children will quickly become disillusioned if the teacher's help is not immediately available. However, techniques can be made clear, even to children not yet fully literate, by a workcard relying on only two sketches and three well-known words. Workcards can also be used to give older or abler pupils facts about the same apparatus (for example, the pitch), suggestions for things to observe, and questions for thought, experiment and research.

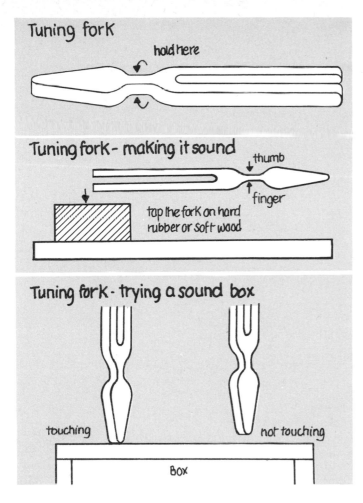

Tuning fork

hold here

Tuning fork - making it sound

thumb

finger

tap the fork on hard rubber or soft wood

Tuning fork - trying a sound box

touching

not touching

Box

Examples   Here are some examples. Either copy them and try them out, on yourself first and then on pupils, or invent some yourself.

Make sure that the language is simple enough because the point is lost if children have to ask what the words mean. Also too much material on each card inhibits the less able and less confident children.

How have you found workcards most successful in practice? What hints have you for other people making them? What, apart from the time it takes to make them, are the problems with using workcards?

What science can the teacher start by looking for?

Experience of things which make noises or notes   Try everything you can think of, with a colleague or maybe two children.

Tapping   Hot-water pipes, the fish-tank table and desk tops, glass coffee jars, milk bottles, empty pottery plant pots, the bottom of an empty drawer, the waste-paper bin, a tray, half a coconut shell, any empty tins.

Shaking   A tin of drawing pins, a jar of peas, a yoghurt pot with small pebbles in the bottom, nails or screws in a bottle.

Class musical instruments   Try these. Some are for blowing and some are for twanging. You will surprise yourself with the amount of material you discover. (Children given this activity to do will surprise you too.)

Noises and notes   From the collection you have tested, make two lists, one of the things which just make a noise, and one of the things which produce a recognizably 'musical' note. The best test for this is to try to sing the note to yourself.

Consider whether this is a useful idea to put to children as well: 'Is this the kind of sound you could sing?' Try it. How does it work with pupils? (See also page 14.)

The difference between a noise and a note is not just playing games, but an important piece of physics, though one does not teach it at this stage. A musical note has one main vibration in it, with a definite number of vibrations per second (see page 16). A noise is a jumble of mixed vibrations.

High notes and low notes   Young children and some older ones may not know at the beginning what 'high' and 'low' mean exactly, and one should certainly not rush into a discussion of the relationships

5

between string lengths and pitch before they are clear about the sounds. Nuffield Junior Science *Teachers' Guides 1* and *2* describe a 'sounds investigation table' for seven-year-olds, saying 'give the children plenty of time to handle the materials'. (See page 4.)

See bibliography: 32, 33.

So what should you look for and listen to before you start working with young pupils?

You could choose a set of instruments to make the point that there are high notes and low notes:

A couple of chime bars, one fairly high pitched and one low pitched, but none in between.

Two recorders, of the highest and lowest pitch available.

Two metal tubes of the same kind hanging from a frame, one long and one short.

Try these for yourself; try other pairs too. The objectives in this exercise are:

To be sure about high notes compared with low notes.

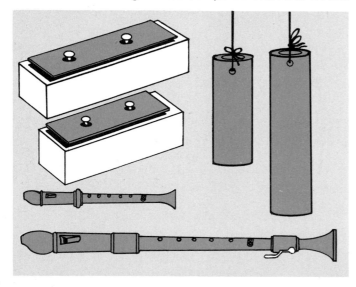

To discover that you can get a high note in several different ways, and a low note in several different ways.

To establish that pitch is distinct from the sound of the instrument.

Analysis Looking back on this small but fundamental set of activities you can see that you have built up:

1 Classes of 'things which make noises' grouped according to the way the noise is made. There are probably four classes, but check this.

2 Two groupings of the same objects, based on the criterion of whether they make a noise or a note.

3 A method of isolating one factor only for clear identification, that is, high or low pitch.

4 A basis for forming the general concept of pitch regardless of instrument.

Notice how these points correspond with some of the logical stages in early modern mathematics. For example, no. 4 above makes a close parallel with the establishment of the concept of number, say the 'threeness of three' independent of apples, bottles or cats.

Now look at these Science 5/13 Objectives for children learning science, for Stage 1:

Familiarity with sources of sound.
The ability to group things consistently according to chosen or given criteria.

Check their relevance.

With children Try the activities you have tested for yourself on some children. They will almost certainly find some things you missed, though you may have to do some guiding too. Results from a small group are usually better than those from an individual, because a child working alone may lack confidence in judging pitch.

2 Guided discovery and vibration

What makes things make a noise or a note?

Finding this out is an excellent activity in which the teacher can help children to make guided discoveries, at the same time learning with them and about them.

The basic investigation   Work with this unorthodox musical instrument in order to escape from preconceived ideas and learnt facts. Stretch a firm rubber band, say 10 cm long and 4 mm wide, over an empty instant coffee jar (8 oz size) across the top and bottom, down the sides.

With a finger, pluck the free bit of rubber band across the top of the jar, in the middle, several times and do three different things:

1   Listen.
2   Watch.
3   Feel, with a fingertip to one side of the middle.

Variables   Consider the experiment very scientifically. Was it only this particular rubber band that would produce the effects you got from nos 1, 2 and 3? Will a different rubber band do the same? Try a thinner one, a longer one, or one of a different colour. Will only one sort of jar work? Could you use a jam-jar, a staffroom coffee mug, an open tin, a cardboard box (such as a chalk-box), a big Sellotape reel, or one's own thumb and fingers, two thumbs, or teeth and one thumb to pull the band tight?

When you do this with children you can encourage them to try as many different ways of stretching the rubber bands tight as they can think of. How is a child to know which variables will have which effect?

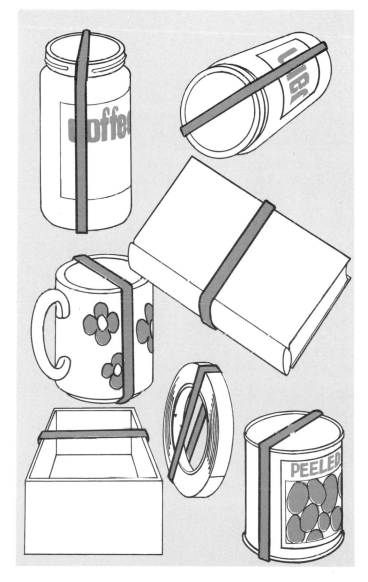

Do the same things happen when you stretch a band round a book? It's not the same musical kind of note, is it? Why not? It's the same rubber band, it's just as tight, but what is different?

As Valerie (six and a half years old) said, 'It can't shiver in the middle.'

So if you put the band back on the coffee jar, pluck it, and quickly stop it shivering in the middle with your finger, what happens to the musical note?

This is how we learn the science of sound.

More tests   Other things must be tried before making a general statement.

A guitar string   What happens when you pluck it? Now slide a book under it so that it touches. Next try touching the string with a finger while it is still moving to see if the results of the first experiment hold for other things.

Do the same test with a zither, if the school has one (see page 22).

Tapping and hitting   What about things that are tapped or hit to make a musical note? Try the following:

A hanging metal tube (one of the tubular bells)
A bar on the glockenspiel
A cymbal
A hand bell

See also page 35.

A bell   Nicholas (seven years old) said: 'Valerie said the rubber band shivered in the middle, but the bell (the old brass school-bell) seems to shiver all round—at least it tickles when I touch it.'

Nicholas went on testing the bell, one strike at a time, holding the edge against his hand, his cheek, his ear and finally his hair. Every time he got the same result.

A comb with tissue paper   Test different kinds of paper before introducing this to children since some are not suitable.

Maria (eight years old) wrote about the comb and tissue paper experiment: 'When you get a piece of tissue round a comb and zum it makes your lips tiggle and it vibrates.'

A kazoo   *Chambers's Twentieth Century Dictionary* calls a kazoo 'a would-be musical instrument'. Get one for yourself—it is very cheap—and experiment to see what you can learn and teach with it. The *Oxford*

English Dictionary says it was invented in Dakota, 1884, 'to give pleasure and satisfaction to the small boy'.

John (aged four and three-quarters) was playing with a kazoo in his nursery infant class: 'Mine goes bzzz. . . .' He said to an adult trying to be appreciative: 'No, not buzz, bzzzz. . . .'

So far he had only made one note, since he had not learnt to hum tunes, but he was making it in quantity, and had grasped an important point.

With the last two 'instruments' you make the note but the tissue paper, round the comb or in the kazoo, responds to your vibrations and adds its own.

The link between vibration and sound

There are plenty of other ways to establish the link between vibration and sound. One very common demonstration of this link described in books is a drum, with rice-grains or puffed wheat jumping about as the drum-head is tapped (see Ladybird *Sounds*, page 10, or Science 5/13 *Early experiences*, page 33).

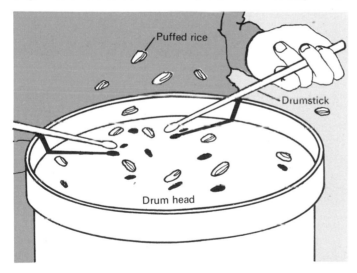
Puffed rice — Drumstick — Drum head

The piano is a good stand-by, as it is for almost all experiences in sound. Some of the links or explanations are too difficult for children actually to discover, and should be left for later.

See bibliography: 4, 9.

In Nuffield Junior Science *Teachers' Guide 2* (pages 62-64) there is a very good description of a class trying this kind of guided discovery. The teacher reported that 'All the class became vibration conscious.' 'The word "vibration" appealed,' and they found it, among many other places, in their own throats.

See bibliography: 33.

You have heard, seen and felt evidence which seems to show that sound is produced by vibrating things. It is not necessary to tell children this; they can be helped to find it out themselves; this guided discovery works. Test it with a small group or a class.

What to do to get notes: blow, bang, twang?

Children might begin by thinking about what they need to do physically to get notes out of instruments. Then they might classify the instruments they know, or their classroom versions, according to these criteria. Here is an opportunity for children to use their knowledge and their ability to organize information and to contribute to class or group activities.

Blowing This is familiar to most children from recorders and whistles, for example. However, they need to think about instruments such as the comb and kazoo, where the activity cannot exactly be described as blowing, since you have to hum with your mouth open.

Banging This includes tapping, hitting, striking. What about shaking, as in rattles, where the action makes one part of the instrument hit another?

What instruments will children know in which these actions happen? Probably several very familiar ones. Would you include the piano?

Twanging The instruments which you pick, pluck, 'ping', from ruler or rubber band to guitar, harp or tea-chest bass, form a favourite and easily recognizable group.

The rest of the orchestra
Somebody will soon realize that the above groupings refer to activities in parts of an orchestra (see Ladybird *Musical Instruments*, bibliography: 3).

This immediately reveals a gap in the classification which should include the largest group of orchestral players; these are not blowing, banging or twanging. How will you name and classify the activity of the first and second violinists, the viola players, the 'cellists, and so on? Guided discovery can happen in areas which are not experimental.

Classification into sets
Children who have learnt about sets in mathematics might try grouping instruments by the method of playing them. Those who know more can make Venn diagrams showing the material of instruments as well.

Where will you put the piano, organ, musical box, electronic organ, Moog synthesizer and human, animal, bird voices?

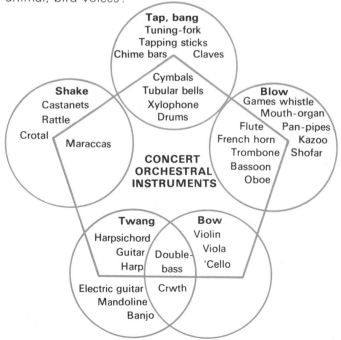

What makes the sound in the instrument?

When children have learned from their own experience that vibration causes sound, they can carry out their own research into *what* it is in an instrument which actually makes the noise, that is, *what* vibrates.

Bells of all kinds
These are easy to start with, for instance, tubular bells, front-door 'ding-dong' chimes, hand-bells (or the old school bell) and the tiny jingle bell, called a crotal, from the Greek word for a rattle. The American plant whose seed-pods are called 'rattle-boxes' belongs to the genus Crotalaria, and the family of Crotalids are the rattlesnakes.

The stringed instruments
These are not difficult to investigate. Test any guitar, zither, violin or banjo: some children start by thinking that the sound of a guitar comes out of the hole in the wooden sound-box or body. Suggest that they look at pictures of banjos and electric guitars which have no holes?

Francesca (ten years old) suggested holding a sheet of paper over the hole in a guitar body; there was still some sound, from the plucked string, but less than with the hole open. This might lead to the investigations on pages 37-39.

The piano
Do the keys make the noise? If you have access to an old piano, or can look inside the school one, a few experimental 'pings' on the strings will be convincing. Then you can look at a zither or a harp (or pictures of them). (See also pages 20, 22.)

Wind instruments
Here is a real problem, since in wind instruments what actually vibrates is the air inside the tube.

The sections of an orchestra may be called woodwind and brass, but it is very hard for children really to comprehend that what vibrates in wind instruments is the air (the 'wind'). For this topic see pages 27-30.

In reed instruments this becomes even more difficult. The best simple demonstration is probably the drinking-straw pipe shown on page 29.

3 Exploring variables

Children want things to do; teachers have objectives for them to achieve. Playing can be very positive so long as it is prevented from deteriorating into aimless playing about. Teachers might use it to explore variables.

The rubber band

You can assemble a rubber-band band in the classroom as follows. You will need:

Rubber bands, especially ones which are about 4 mm wide and 9-10 cm long. (Some postmen use a slightly wider, slightly shorter variety round bundles of letters.)
Empty jars, especially coffee jars.
Pottery coffee mugs (not the thin plastic or cardboard ones).
Rigid empty boxes and tins of all sorts, from the small greengrocers' wooden fruit trays to empty smooth-edged baked-bean tins.

Put the rubber bands round the boxes and play (see page 12), plucking—'pinging'—the bands.

What variables can be investigated?

1 The width of the rubber band
Use two or more bands of the same unstretched length, but of as different widths as possible, round the same 'sound-box'. Stretch them as equally as possible, thus making the tests 'fair'.

2 The tightness of the rubber band
Slacken and tighten the same band several times on one sound-box, testing the pitch after each adjustment.

3 The length of vibrating section
(This is the plucked part of the rubber band.) Tests can easily be set up by putting a pencil under the rubber band across the top of a wide-mouthed jar, and plucking and testing the section of the band on each side of the pencil separately. If the pencil is moved steadily across the jar, plucking the band meanwhile, a range of some two octaves can be produced.

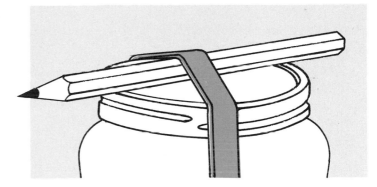

The method will quickly be discovered. Pulling the band a little more tightly over the opening will give a higher note, and the band usually provides enough friction to stay tight. Conversely, loosening it will give a lower note. Thus tightening or loosening a string is the basic method of changing its pitch. This will become clear.

Playing the band
A group of children in a semi-circle playing a set of bands and jars can produce very satisfactory melodies. A wooden fruit tray with long rubber bands round it makes a crude imitation of the medieval rebeck, or the Welsh *crwth* (pronounced 'crooth') which had two plucked strings and four bowed strings, combining the principles of the harp and violin.

Analysis

What science can the teacher expect children to gain from this exploration? It provides an experience of fair testing which is basic to all scientific method, that is the isolation of variables or testing one thing at a time. It also specifically gives pupils three factors influencing the pitch of the note from a stretched string.

From no. 1 above: the thickness ('weight' or mass per centimetre) of the string (band).

From no. 2 above: the tightness (tension) of the string.

From no. 3 above: the length of the vibrating part; the pencil may make the band slightly tighter as well.

Getting a high note

When you test these three variables, how do you find you get a higher note? With a thin or thick band? With a tight or slack band? With a short or long vibrating length?

You will probably have known all these answers from the beginning, but test just the same, to see how a child finds out. Then do it with children.

Mnemonics

Can you invent any phrases or mnemonics to help children to remember which conditions produce what sort of pitch? For instance, phrases like 'long and low' (these are not for direct teaching, of course)? They are particularly valuable when you produce them just after a child has made a discovery; then the point sticks.

Scientific thought is often directed towards looking for regularities and patterns in events or facts. Some children will be much quicker at this than others; can you get the quick ones to show the slower ones?

What about this little triangle?

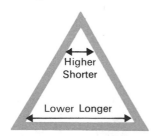

Hollow or solid? Further variables to test

'Empty vessels make the most sound.' Is this true? When children come across folklore which can be tested, they can use the opportunity for scientific activity. Encourage them to think about such things.

What can the teacher offer them for testing this saying? What do they suggest themselves? Do they know its figurative meaning?

Discussion will stimulate plenty of ideas, and the more things the children test, the sounder the basis for forming a generalization: 'Yes, it is true/no, it is not true/it is sometimes true but not always (the exceptions being . . .)'

Investigations   What could be tried?

Two coconuts   One should be complete; the other one should be halved and emptied, and then its two halves should be stuck together again. (If Copydex is used, the halves can later be separated.)

Solid blocks of wood   These should be the same size as the hollow percussion-band 'blocks'.

Somebody should ask if they are both made of the same kind of wood. If anyone does ask this question, your science teaching has been successful. The children have learnt to isolate variables, and to test only one at a time. This comes quite late in a child's development, often at the age of eleven or twelve, or later.

Coffee jars   Test two, with their lids on to avoid splashes. One should be full of water and another one should be full only of air.

A box   It should be alternately filled (say with a dry towel) and empty. Check the different sounds you hear when you tap it when it is full and when it is empty. Repeat this a few times.

A tin   full of sand or sawdust or stuffed with the board-duster, and empty.

There will be no lack of suggestions (page 19).

Generalization   The results will enable the children to understand why a double-bass or cello has a huge wooden box. They might then be able to appreciate the difference between a Spanish-type guitar and the guitar with a plug-in lead played by pop stars equipped with electric amplifiers.

Scientific method involves observing, guessing, testing fairly, modifying the guess if test results say you should, then testing more widely, and generalizing when you have enough results. Most children can carry out these processes when they are attracted to the subject matter and feel that their activities are appreciated.

Do you agree with T. H. Huxley's remark: 'Science is, I believe, nothing but trained and organized commonsense!'? How can teachers best help pupils to train and organize what commonsense they have? Does this help them to develop?

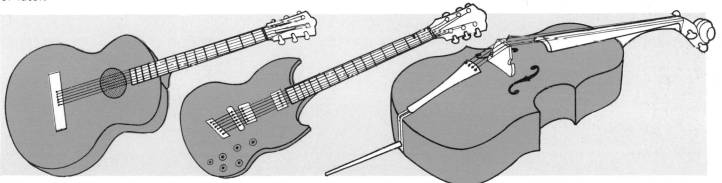

4 Tuning

When children have succeeded in making a musical note they will need to learn how to get it to the right pitch, especially if it is to be part of a tune.

Some children will immediately 'know' a high note, a low note, or a middle note, and will be able to decide if the note their instrument is producing is the right one or not.

Some will recognize the pitch more easily if they are used to singing to themselves the notes they play.

Andrea (aged nine) played regularly on a two-octave classroom glockenspiel. The teacher heard her playing a melody which had one high note, which she left out each time it should have occurred.

Teacher: 'Why don't you play the top note? There's the bar that makes it.'
Andrea: 'I can't sing so high.'

She did not look as though she was singing at all; but evidently she was doing so internally, like the silent 'speaking' of words in reading or writing which many people do when, for instance, addressing envelopes.

Do you sing a melody to yourself, aloud or silently, as you pick it out on an unfamiliar instrument?

Do you think that doing so helps children to recognize the exact pitch, for instance, for tuning?

Does this suggest any special choice of instruments for beginners?

Do you or your pupils think, therefore, that one has to tune one's vocal cords all the time while talking?

Tapped instruments

Many instruments played by tapping have a fixed pitch, like bells, though if these are struck too hard or in the wrong place they will be out of tune.

Other instruments, such as coffee jars of different sizes, can be tuned by adding water (see also page 15). Here is a real experiment, starting with one of the first enquiries on which children base their scientific work, namely, 'What happens if . . .?' Suggest first tapping the empty 8 oz, 4 oz and 2 oz jars, and then maybe two or three 4 oz jars, and then one 8 oz jar with some water in it, and so on.

It is important to test this yourself, since memories of a different experiment which has opposite results, that is blowing across the top of milk bottles, may confuse you. (See also page 27.)

Help children to test experiments with an open mind.

For your own information, tapping gives the lowest note when the jar is full of water because this consists of the largest mass of glass and water to vibrate. But try it; also try tapping regularly while you pour water steadily into the jar; listen to the changes in pitch.

Do you find any mathematical connections, for instance, when the jar is about half full? The problem is not simple to solve, since both glass and water are involved. Try jars or a set of exactly similar bottles, such as milk bottles, to see if you can make an octave. This might lead to the investigations on pages 27-29.

Children and octaves

Almost all students and teachers recognize an octave when they hear one. How did you learn yourself? Can you sing or hum two notes an octave apart, going up and down?

Try this with children. Can they copy you? Can they do it at another pitch? Are there any who cannot do it? If so, what is the problem?

Recognizing the pitch of sounds and their relationship?

Controlling the vocal cords?

Vocabulary, the word 'octave'?

Work on these with children. Maybe they can copy from you and recognize the related pitch values without knowing what the interval is called. Can you find out?

Recognizing an octave   What everyday experiences can you use to help children who do not know an octave, and to reinforce the knowledge of those who do?

Popular tunes   Are there any popular tunes with octave intervals in them? 'Somewhere over the rainbow' is one. How can you introduce children who have no access to a piano to the invaluable visual experience of octaves which the piano or organ keyboard provides? A piano accordion is appropriate.

A mouth-organ   Experiment with a mouth-organ. (Only its owner should play it, for hygienic reasons.) Most mouth-organs have a complete octave in the centre of the range, and possibly a whole octave below or above as well. One pupil might start by playing the basic central key-note of the mouth-organ and then slowly play a 'scale' going up; the children should stop him when he gets to exactly one octave above it. (See also page 16.)

The children   Which are best trained to do this kind of exercise? Perhaps the choirboys, children who have piano lessons, or recorder players. Test this with them. Can you get any picture of the differences made by early training and experience?

If you have an opportunity of testing octave recognition with children of widely different ages, can you do an experiment on Piaget's lines to see if there seem to be developmental stages in your limited sample? That is, is the recognition linked with age groups? Tests could be arranged to find out who hears an octave when it happens and who knows what the word 'octave' means in sound.

Working with octaves
Playing octaves   Once the interval is familiar, children can very much enjoy 'going up' and 'going down' by octaves.

The piano is the best instrument for this. Why is it useful for the children to start with a pitch they can themselves sing?

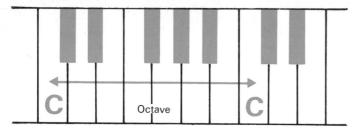

Counting an octave Counting the 'steps' inside one octave (singing, playing or counting white piano keys from C to C) may give children the clue they need to understand the meaning of the word. It is connected, for example, with the words 'octagon' and 'octopus'. Knowledge here will depend on the vocabulary the children may have acquired from other work.

Note that one octave includes both ends of the scale, so two octaves consist of only fifteen notes, three octaves only twenty-two.

Harmony with octaves Children may have detected this much earlier in their study of sounds. Those who play instruments can help here. All are likely to learn more about it by using experimental methods to discover how to achieve the interval of an octave, for example by halving the length of the part of a guitar string being plucked.

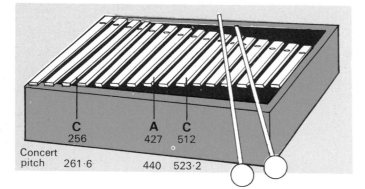

More advanced work

The numbers of vibrations per second are marked on tuning-forks (see page 36) and glockenspiel bars. Look for the numbers on two tuning-forks which have an octave interval between them. An older child who sees 261·6 on middle C and 523·2 on C one octave higher will realize what is happening mathematically, without necessarily realizing that the figures represent the number of vibrations per second.

Can you produce at least one piece of evidence for your pupils?

Can they predict what number of vibrations per second will be given on the glockenspiel bar or tuning-fork one octave below middle C? Can they find one to check with? (A music shop will help here.)

Children may ask why these are not whole numbers. The answer is that the number of vibrations per second is fixed for any pitch, but people have decided what pitch shall be called, for example, middle C. The note A above middle C which is used to tune violins and the other instruments of the orchestra, has been agreed internationally as the tuning note. This is concert pitch. It has a frequency of 440 vibrations per second or 440 hertz (shortened to 440 Hz). The other notes have to fit in so that the sound intervals are correct; this leads to figures such as the one for 'middle C', 261·6. Both Science 5/13 *Metals* and *Like and unlike* give the figure 256 (hertz, or vibrations per second) for 'middle C', which gives 512 Hz for the C above middle C.

The physics of octaves will be more important later on.

See bibliography: 11, 12.

5 The guitar in the classroom: making and testing hypotheses

The ordinary guitar is probably the most popular instrument among pupils, and may well be the most available stringed instrument in the classroom.

Julia (aged eight and a half): 'A guitar is a string instrument and also a musical instrument. You can get lots of notes out of six strings.'

Tuning the guitar

Investigations

Choose one string. To help less confident children to remember which one out of the six, you could tie a short scrap of ribbon round it at one end. Now set up a finding out session.

1 Where do you tune it? (At the top, at the peg. . . .)

How do you tune it? (By tightening or loosening the string; by turning the peg.)

The children will tell you.

2 What do you see when you pluck and let go?

Sarah (seven): 'It looks furry—fuzzy—when you let go.' Link this up with vibration (page 9).

3 Play a note. While doing so, let children put their fingertips on the sound-box, on the sides of the finger-board, on other strings. Can they feel anything?

Organize things so that there are not too many children round the guitar at once. Three is probably a good number, one trying each position several times. Note that scientific method involves checking each result.

4 What are the frets for? These are the ridges across the finger-board. What do they do? This depends on what *you* do, of course. Again, children will know at least part of it, but need to be encouraged to think accurately.

5 What happens when you move your finger from fret to fret, plucking at each position? Start on the fret which gives the longest length of string and work step by step towards the sound-box. (Stay on the same string.)

There are two different things here: the length of the string and the pitch of the note. (Most children will only think of one at first.)

6 How many frets are there? How many different notes can you make with only one string? Guess your own answer first; then count. Was it more than you guessed? Try this with pupils. Then get them thinking about this question: if you can get so many notes with only one string, why do you need six strings?

7 Put your finger behind, say, fret no. 8 counting downwards, and pluck the bottom part of the string. Then pluck the part above your finger (that is, the part nearer the tuning-pegs). Does the result agree with the generalization you worked out earlier about length and pitch?

8   Can you, without actually saying anything, get a child to propose putting a finger exactly halfway along the string? You might get this to happen by edging your finger along the string, or encouraging a child to do so. Move your finger nearer and nearer the halfway mark until either the symmetry or the special pitch relationship of the two halves occurs to someone. Do both halves make the same note?

Is this a general hypothesis? Choose any string, measure its length from the bridge on the sound-box to the ridge near the pegs, divide by two, put your finger on the halfway mark, and pluck above and below to test the hypothesis.

Why measure? There are at least two reasons why you cannot count frets as if they were centimetres. Look at the guitar and see.

9   Now play in turn the open string, the top half of the string, the bottom half, and the whole (open) string again. What are the musical relationships of the notes you get? (Low, high, high, low, with an octave interval between your low and high.) The vibration of the whole string gives a different note from that of the half-string. Can you *see* any difference?

10   Try plucking the whole string and watching its middle closely. Perhaps you could use a large magnifying glass such as an Osmiroid Magnispector? What can you see if you pluck the half-string, and even shorter lengths? Can you see any difference in the rate of vibration? Which is slower? Which faster?

Check once more with the whole and half-lengths (see page 11). You should be able to make a good guess (a hypothesis) and, after several observations, a generalization from these investigations (see page 20).

How much of this rather concentrated observing and thinking does a child of say eight grasp the first time? Is it a matter of readiness?

How far do you think it valuable to let pupils try out such ideas (a) in play *before* you do it with them, or (b) for fun or for reassurance *after* you have done it with them? Don't they always want to get hold of the guitar themselves? And shouldn't they?

The guitar string and effective variables

Choose one string and try to make as many different sounds, not only notes, as possible.

How many ways can you think of? You might try:

Tightening the string.
Slackening the string.
Holding your finger on it part way down.
Plucking the string very gently.
Putting a soft cloth in the sound-hole.

Get pupils to do these activities with you.

Every factor you change which affects the sound from that one guitar string is an effective variable. Make a list of them, and the effect which each has. Get pupils to do this with you in a class discussion.

The items will fall into three groups:

1   Those which make the sound higher or lower, that is, affect the pitch.

2   Those which affect the volume, making the sound louder or softer.

3   Those which make the sound different in another way, changing its quality, perhaps to 'dull' or 'tinny'.

This does not take long, but is a good exercise in listening, observing and in scientific thinking.

Generalization   What is the general effect of

shortening the plucked part of the string?

This can be tested on the stringed instruments normally played with a bow: violin, viola, 'cello, bass, double-bass, or on a tea-chest bass (if you can find one). Does the generalization hold?

Let a pupil test on a violin the sound made by plucking the short length of string between the bridge and the 'tail-piece', first predicting the pitch.

Extensions   The hypothesis can be extended to other investigations.

A harp   Look at a picture of a harp, or at a real one. Where do you expect to get high notes? And where the lowest notes? Why? Get children working on this; check in every way you can, for example, with a piano (see below). (See also page 22.)

A zither   Much can be learnt from this instrument. Look at one and again predict where you would expect, and get, high and low notes. (See page 22.)

A piano   Look inside a piano, or at the outside of a grand piano. The shape of the latter's case may tell you what you need to know. Which end of the keyboard gives the high notes and which side of the case has the short strings inside it? Compare this with the harp. (See also page 22.)

6 One string and many strings

The monochord

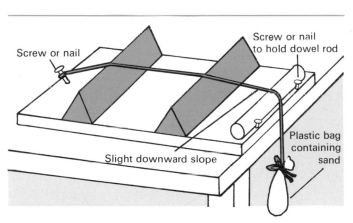

Screw or nail
Screw or nail to hold dowel rod
Slight downward slope
Plastic bag containing sand

The Greeks played monochords 2500 years ago. There is a French one with a tuning-peg in the Victoria and Albert Museum which is nearly 2 m long, and was invented in 1883.

Making a monochord   This simple apparatus, with a string over a board, is very useful and many discoveries can be made with it.

See bibliography: 2, 11, 34.

Make sure that the nail or screw is firmly in. Violin string or guitar nylon is safer than piano wire, which is sharp-tipped and springs back. The end of the board with the hanging string must be kept just on the table, or the whole thing may tip up.

Discoveries to be made
Changing the note   The connection between the pitch of a note and the length of the part of the string which is actually vibrating.

What happens when you increase the tension? Hang a heavier bag of sand on the end of the string. This will make the tension on the string greater. This compares exactly with tightening the tuning-peg on a guitar. The variable can be measured, say by doubling the load.

Volume   What happens when you pluck more vigorously? This leads to a new experiment of changing the volume—the loudness—of the sound, and observing what makes the sound loud or soft.

The string can be seen to be vibrating more widely when it is making a loud sound than when it is making a soft one.

A loud sound can be a high note or a low note, though many children may not be clear about this. They may say that raising the pitch of a note is making it louder. Do they hear things differently from adults, or is it a language problem? You can test this, and can help them to differentiate.

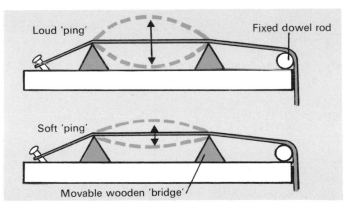

Loud 'ping'
Fixed dowel rod
Soft 'ping'
Movable wooden 'bridge'

For many children volume is simply a word on or near a knob on the TV set or transistor radio. Many of them will never know what it is for or use it even if they do know. How can you help them to appreciate volume of sound and changes in volume?

Stopping the sound
This is done by touching a vibrating string and is called damping. Children enjoy plucking a string and then at once touching it with a finger. In this way they gain experience of vibration itself, and of the intentional stopping of the vibration, and therefore of the sound, which musicians often use.

Can they find out on what instruments this is practised? They can try bells, which vibrate for some time after being struck. They could also try the soft pedal on a piano and look inside to see what is happening.

Instruments with several strings

The zither
This instrument is very good for learning the science of sounds from strings. It can easily be tuned and the high notes have shorter strings than the low notes. Most zithers have a few thicker, heavier strings giving notes an octave lower still. (See also page 20.) Children can wonder and find out on it.

The harp
It is unlikely that children will have an opportunity of playing a real harp. They may, however, see one on TV or in a museum. There are pictures and posters to be seen which show this instrument, for instance the Guinness harp which has only nine strings. (See also page 20.)

What does the harp have to offer? Its most important characteristic is the enormous and visible difference in length between the strings which give high notes and those which give low ones. This is a special opportunity for children to test a generalization or hypothesis. Let them do the guessing and checking. Which end is which? Which hand plays the high notes?

The inside of a piano
This is informative, but it is too complicated for many children, partly because of the diagonal arrangement of some of the strings and the 'doubles', but also because of the mechanical steps involved in pressing a key in one place and getting a string note from another. (See also page 20.)

However, the piano keyboard is useful because it can be compared with the early mathematics 'number line'. Going up or down and hopping over six white keys (notes) each time, provide training in the recognition of the octave and the octave principle. (See also pages 15-16.)

Rubber-band versions
These are much appreciated, but are tricky because several variables are involved and the results obtained with them by children may even be the complete opposite of those expected, and therefore useless for simple generalization. For instance, simple stretching makes bands thinner as well as tighter and longer; the first two factors should raise the pitch, but the third tends to lower it.

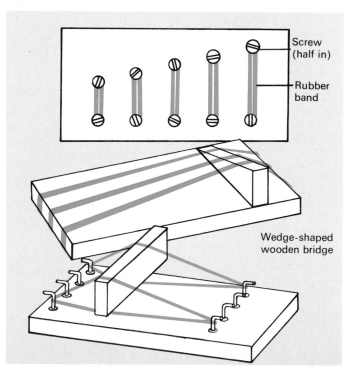

Screw (half in)

Rubber band

Wedge-shaped wooden bridge

Test the rubber band instruments illustrated here, first in your mind and then in practice. Can you see advantages and disadvantages about each of these? Which do you think, and which do you find, the most satisfactory in helping towards a real understanding of vibrating string instruments?

The violin   As Julia (aged eight) said: 'A violin is the same as a guitar but you have a bow to make the music.'

Compare a guitar and a violin. What things about them are alike? This is an interesting exercise,

especially as one can play the violin by plucking the strings (a technique called pizzicato, from the Italian word *pizzicare*, to pluck). Tuning is similar: guitars are tuned to E, A, D, G, B, E, and violins to G, D, A, E.

What are the main differences between a guitar and a violin?

You pluck a guitar to make the strings vibrate; you bow a violin. The effect is different, though the strings vibrate in both.

There are frets on a guitar to give you the finger positions for the notes, but on a violin you have no help. Try.

A classroom music collage (see page 26)

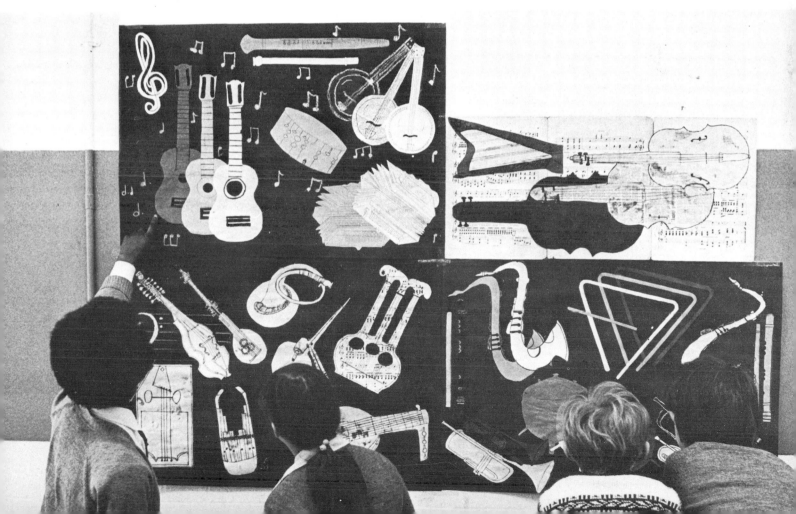

7 Explanations

Explaining observations

Many children, especially young ones, are happy to accept that a variable has a particular effect. To them this is a sufficient explanation in itself.

How, for instance, do children explain, to you or themselves, what different factors actually make the notes from plucked strings high or low? Can they do it at all? Perhaps all we can say is that we happen to hear rapid vibrations as high notes, and slow ones as low notes. Children can work out explanations to say why short tight strings vibrate rapidly. Bruce, ten and a half, said: 'The tighter the string, the more it's pulled back after you've pinged it' and 'the short string hasn't got so much string to move to and fro as the long one.'

Some children are able to extend their explanations much further. They may say 'It's like when you . . .' You would be very satisfied if they explained how a glockenspiel works by saying that the longer the glockenspiel bar, the lower the note, that it is 'like when you pluck a longer piece of the guitar string'.

The use of one situation to help explain another serves to underline the importance of providing a variety of experiences. Indeed, 'There is no substitute for first-hand concrete experience' (Nuffield Junior Science *Teachers' Guide 1*. See bibliography: 32.)

When the children have had many experiences of vibrating things producing sounds of different pitch, it is interesting to discuss explanations with them. It is disastrous to push them for explanations too soon, since they will flounder around in words, trying to find the phrase which will gain adult approval, often contradicting themselves in the process.

Testing explanations

Here are some investigations which will help children to test their explanations of vibration.

Tapping sticks   Get one or two thin bamboo garden canes. These canes are cheap, easily sawn into lengths with a junior hacksaw, and the rough edges can be smoothed with a nail-file.

Cut two pieces, as alike as possible, and about 15 cm long. Cut two more pieces exactly twice as long. All of them should be as nearly the same thickness as you can manage.

Use them in equal pairs as tapping-sticks (rhythm sticks or claves). Listen to the pitch of the note the short pair makes, and compare it with the note of the longer pair.

What do you notice? How do you explain it? Do children hear the difference at once, or do you need to drop a hint?

A ruler   Take a 30-cm (12-in) ruler, hold part of it firmly down close to the edge of the table, and twang the projecting end.

Move the ruler to and fro, so that a long end sticks out beyond the table edge, then a short end. Twang it in various positions. Finally slide the ruler from the long end out position to the short end out position. Twang it at each position and listen to the note.

Can you explain the different pitch at the various vibrating lengths? How many other pieces of apparatus or musical instruments can you think of for which the same kind of explanation would apply?

A hacksaw blade   When you look at stringed instruments you will notice that the deeper notes of the 'cello double-bass, piano and harp are produced from thicker strings than the high notes, or even from strings with a tight coil wrapped round them. However, you cannot test such strings to see what note they would make if they were thinner or lighter.

One explanation might be that the heavier the material the more slowly it moves to and fro. This could be tested with something which can be loaded.

Take a large-sized hacksaw blade and fix it tightly with a G-cramp, so that most of it projects from the table edge. Twang it, then twist wire round the end and through the hole for safety; then twang again.

Another method would be to take the longest ruler available and clamp it to the table edge with a G-cramp, with the longer end projecting. Twang it several times to be sure of its note. Load the end by clipping on one or two Bulldog clips. Twang again.

Does the pitch of this note help with your explanation?

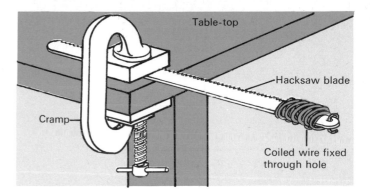

Table-top

Hacksaw blade

Cramp

Coiled wire fixed through hole

Is logic involved here? 'Logical consequences are the beacons of wise men.' (T. H. Huxley) How much in this work on science from musical instruments can you try to help children to think logically? The concept 'Because this . . . then that follows,' is an objective worth keeping in mind even in very simple activities.

What do your pupils volunteer as explanations?

Listen to one child explaining musical processes to one or two others, a good and useful classroom device; you may be agreeably surprised.

Vocabulary and explanation

The problem hindering the child who cannot easily explain an experience may be simply a lack of the right vocabulary. It is difficult for the teacher to know where the hitch comes, but help with vocabulary may overcome it.

Accuracy in the use of words is essential for science and scientists. Musical sounds provide much descriptive vocabulary for young children. Later on exact measurement will need to be stated in words.

Spoken language connects music with science and linguistics. It often matches sound with sound, as in 'cuckoo', the word used in the *Concise Oxford Dictionary* to define onomatopoeia.

Some oral expression does not depend on writing. In fact, some cannot be written, for instance Gaelic mouth music, which is wordless but articulated, humming, whistling and songs without words. Children can and should have some experience of all three, each of which uses a different basic method of sound production.

Thinking, feeling and testing these are scientific activities, as well as fun and culture.

Collecting, writing and classifying sound words
Make a list of words connected with sounds, for example buzz, ding-dong, shriek, ping, bang, zoom. Some of these clearly indicate vibration, while others suggest high or low pitch.

Sort out those you have collected into lists. Here are some suggestions.

Vibration	High notes	Low notes	Others
buzz	tinkle	boom	crash
hiss	whistle	thunder	bang
shiver	squeak	growl	zoom

Look at the actual sounds in the first three lists. Do you see certain sounds which seem characteristic of each list, for instance, 'i' or 'e' in the high notes column? Test this hypothesis. This is a scientific exercise.

Which part of the following two-sound words do you expect to have the higher pitch?

Ding-dong (like a front-door bell), click-clack plink-plonk, jingle-jangle. (See also page 25.)

Does your hypothesis work?

Of the pair 'tick-tock' and 'ting-tang', which has a metallic and which a wooden sound? Take 'ting-tong' and 'ding-dong': which has the 'tinny' sound?

Help children to widen their powers of expression by collecting and classifying sound words and making lists. Encourage them to practise sensitive listening to sounds and the words which go with them.

Listen yourself to the children's spontaneous sound words. They might be stimulated to try inventing some. And you may learn what things which are familiar to you sound like if you have never heard them before.

Words and instruments
As experience of making music accumulates, so a wider vocabulary is needed for at least two other purposes:

To describe what you do to the musical instrument to get a required or satisfactory result.

To name the instrument, if only for communication, and to tell other people about the sounds it makes and to understand them (for example, in books and scores).

Neither of these sets of vocabulary items is essential at early stages. The apparatus will make the same sounds for you, whether you can name it or not, and whether or not you have the right word for your activity. The sets will, however, be useful in later work.

Identification
All the instruments known to the children can be given their correct names. A few children may want to make personal music scrapbooks or a class collage. Local pop and classical music shops will help with catalogues and folders, and some children will become very enthusiastic and knowledgeable about the differences between types of string, range of pitch and methods of tuning. These details have their scientific importance too.

See bibliography: 3.

What information do you find you can collect from a class of children about the names of orchestral instruments and the ways in which the sounds are made? What about the more limited range of pop group instruments? Can you help them to link the two?

8 Wind instruments

This topic is difficult, because while you can *see* a string vibrate, it is much harder to be sure that you know what happens when you make *air* vibrate. Some of the complex facts about brass wind instruments are not yet fully understood.

However, some wind instruments are easy to play and every school has them, so let children use them and find out what they can, which is quite a lot.

The bottle organ

Many books describe and illustrate the method of blowing across the tops of milk bottles containing different volumes of water. The important factor is the difference in the volume, or rather the height, of the *air* in the bottle, since this is what will vibrate. The teacher needs to think hard here, because tapping the same bottles, which is easier (and quieter), gives the opposite results. It is the depth of the water which decides the pitch, so the more water in the bottle the lower the note when tapped but the higher when blown, because there is less room for air.

Simple tubes

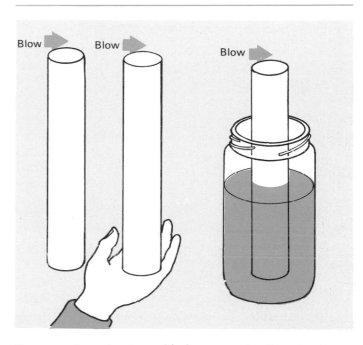

Get a cardboard tube; a kitchen-towel roll centre is very suitable.

1   Blow across the top until you find the right position for the 'note' it makes. Once you have found it yourself, you can help children who have less expertise.

2   Put your hand under the bottom end of the tube and close it while you blow. Did you expect any change in pitch? Did you hear one? Did you expect what you got? Test this again, more than once.

3 Fill a tall jar (say 8 oz instant coffee size) with water. Hold your cardboard tube so that the bottom end just dips into the water, that is, the tube is closed by water. Blow across the top again. Move the tube down until it is roughly half full of water, then till it is mostly under water; keep testing the note by blowing.

4 Blow as you move the tube both upwards out of the water and then down again. What happens to the pitch of the note?

What instrument are you copying? The trombone. You could make a model which will be more like a trombone, using just the tube (barrel) and piston of a bicycle pump. The Swanee whistle works like this too.

See bibliography: 29, 38.

Explanation Much of what is observed in this tube and water experiment you can easily explain. No. 2 is difficult, and for this you need a physics textbook.

See bibliography: 34.

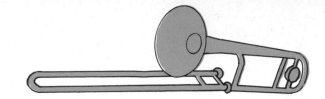

Children sometimes find it difficult to get a note at all by blowing across a tube. Do you find you help them best by showing them how, or by encouraging them to go on trying until they hit on it?

Predicting and testing Try a simple exercise in prediction, on the basis of the hypothesis that the length of the air inside a tube affects the pitch of the note you get when you blow across the top of the tube and make a note. You know what effect a long tube or a short tube is likely to have.

Collect all the ball-point and felt-tip pen caps you can find. Arrange them in order of length, and see if your hypothesis works by blowing across the open ends and listening to the pitch of each note.

Let children try with their own. Some will have a set of several pen-caps all alike. They can predict for and test these too.

Does it work for longer tubes, for example the barrels of old pens or the Biro refill containers?

Paper straws for testing hypotheses

For this you need a packet of waxed paper straws. (The coloured plastic ones are not suitable.) These straws may be difficult to find, but may be provided for school milk. They are sometimes called 'Sweetheart' straws. You will also need scissors.

1 Flatten one end of a straw and cut off the flattened sides (about 1 cm long) with scissors. Put the cut end just into your mouth, not between your lips, leave the end free inside and blow. You may not get the squawk at first, but it will come.

Now predict what you would expect to happen to the 'note' if the straw were shorter.

2 Start squawking again, and while you are doing so, snip a bit off the far end of the straw with scissors. Did your hypothesis work?

If you do not want to risk using scissors while blowing, make a set of straws of different lengths and play them separately. You will get the same test of your hypothesis—but less fun.

See bibliography: 9, 24.

What does this exercise tell you about the relationship of the pitch to the length of the vibrating air column? Does this support your generalization?

Workcards This kind of activity is very well suited to workcards. It is easy to explain, particularly if you illustrate the cards, and children can get on with it while you are busy with more difficult practical exercises or talking with pupils who need you.

Look back in this book and see how much of the work can be done using workcards. It probably makes for flexibility. Would you agree?

Recorders

These are the most available wind instruments.

Compare the lengths and pitch of recorders of different sizes, also the names of the 'voices'. See the Ladybird book *Musical Instruments*.

Compare the pitch of individual notes on any one recorder with the position of the open hole when the note sounds. Make a table of notes and lengths from mouth-piece hole to finger-hole.

Guessing, that is making hypotheses, is easy at this stage and testing should be satisfactory. The Ladybird book is most helpful in giving the range of notes for each instrument compared with the piano keyboard.

See bibliography: 3.

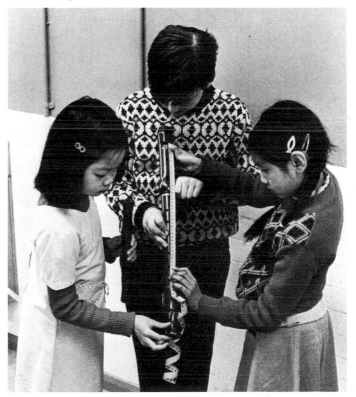

The mouth-organ

A mouth-organ with one side cover missing makes an excellent instrument for study. Air is blown in, and drawn out, but the notes are produced by thin strips of brass of different lengths or widths vibrating because of the air.

Classification into sequence sets

There are many mathematical ideas which one can illustrate using musical instruments. Here is an easily grasped but very significant example which is appropriate when children have learnt something about string and wind instruments. Make the two tables of instruments, and then compare the tables with one another.

Can you classify *these* instruments?

An ordered series of musical instruments by size

Smallest			→	Largest
1 Violin	Viola	'Cello		Double-bass
2 Descant recorder	Treble recorder	Tenor recorder		Bass recorder
3 Piccolo	Flute	Oboe	Clarinet	Bassoon

An ordered series of musical instruments by pitch range

Highest			→	Lowest
1 Violin	Viola	'Cello		Double-bass
2 Descant recorder	Treble recorder	Tenor recorder		Bass recorder
3 Piccolo	Flute	Oboe	Clarinet	Bassoon

For the information and for further ideas see Science 5/13 *Like and unlike*, pages 43, 44, and the Ladybird book *Musical Instruments*. (See bibliography: 3, 11.)

9 Tapped and struck instruments

Bells

The physics of bells is very obscure, since the metal and the shape affect both the pitch and the quality of the sound. This is complicated by the fact that bells often have at least two notes, the strike note and the lower, longer-lasting hum note. Children are interested in the note made by Big Ben, and perhaps they may hear Britain's largest bell, weighing just under seventeen tons, in St Paul's Cathedral. They can do some sound research on the notes of local church bells and clock chimes; Big Ben they can hear from the BBC.

Tubular bells bought or made at school, give pleasure and melody to children's music. *Metals*, Stages 1 and 2 (page 19), describes how to make them.

Copper tubing, the sort used in modern hot water systems, is good for this. The tubing can be sawn with an ordinary or junior hacksaw and filed down.

See bibliography: 12.

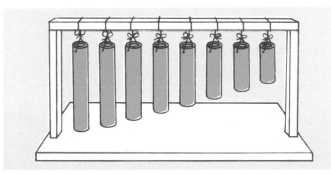

Coffee jars, jam-jars, bottles These make a cheap substitute for copper tubing, so that more children can work on the investigation at one time. Instant coffee jars, all of exactly the same pattern and in 8 oz, 4 oz and 2 oz sizes and without lids, give almost accurate musical intervals of thirds. The jars can be tuned by adding a little water at a time to them and testing the note by tapping (see also page 14).

See also Science 5/13 *Science from toys* (bibliography: 14).

Flowerpot bells Made of real pot (not plastic), these have a fine sound, but are difficult to buy. To get a set which is in tune you need a large number to choose from; some will crack when they are tapped too hard.

See bibliography: 9 (page 35).

Tin-can bells are usually unsatisfactory, though it might be worth trying them.

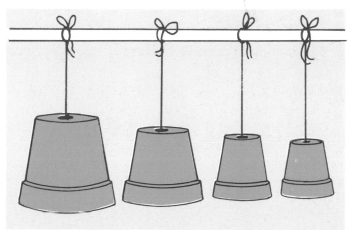

Glockenspiels, chime bars, xylophones

All these depend on the vibration of a stiff rod.

Some books, such as *Science from toys* (page 13) and *Working with wood*, Stages 1 and 2 (page 16) suggest making a xylophone. This is an attractive idea, but it is difficult for any but the skilled to carry out in practice, since some woods do not give a clear satisfying note, and tuning is tricky.

Hugh Garnett's *Practical Music Making with Juniors* describes exactly how to make a small glockenspiel, called from its main component a 'nailophone'.

A two-note xylophone   R. Roberts in *Musical Instruments Made to be Played* suggests making this out of two wooden bars, which give results which are more immediately useful. Here are his instructions for making the instrument, which is held by a cord and struck with a beater. No frame is needed. The timber is Columbian pine planed to a section 2·5 × 1·5 cm (1 × $\frac{5}{8}$ in). The narrow grain should show on the wider surface if possible. The approximate lengths should be 35 cm (13$\frac{3}{4}$ in) for the note d' and 31 cm (12$\frac{1}{4}$ in) for the note g'. These wide shallow pieces give a deeper pitch than wood with a square section.

The battens are tuned to pitch by cutting off small amounts of wood, then they are looped into a pair by a cord and small staples. Suspension points are at the

nodes, at quarter and three-quarter length. The loop gives a safe grip for the left hand. The right hand holds the conventional hard-headed beater.

See bibliography: 14, 17, 24, 28, 35.

Percussion

Drums   Try:

Large and small ones.
Tight and slack drum-heads.
Hitting hard and gently.

Observe, compare and explain the high and low notes, also loud and soft sounds produced in predictable ways.

Safety   Children sometimes suggest rubber balls as drumstick tips. Golf-balls, however, should not be spiked as this might be dangerous.

Other percussion instruments
Cymbals
Gongs
Saucepan and dustbin lids
Suspended half-coconut shells

Safety   Suspended glass jars are not very safe. Children find it very difficult to tie string tightly enough round the necks of jars. (Sellotape should be substituted, although it may come unstuck when wet.)

Do they find out for themselves the damping effect of standing the saucepan lid on the table, as compared with the note from the same lid hanging up? Can they explain why contact with the table has this effect? See pages 22, 35.

Pupils will be able to suggest many objects which can suitably be used in this kind of music making. Can you help them to hear the main note made by each object, for example the half-coconut shell? Perhaps they can sing the note?

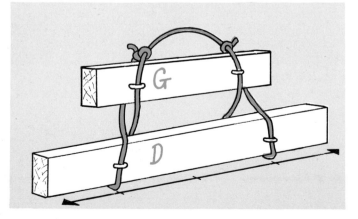

Jingles and tappers

Inventiveness is a scientific and technological asset consisting of thinking of something which might work, and then testing it to see if it does. The teacher can encourage it by suggesting competitions between groups of children. A good subject would be making jingles and tappers. This could include every pupil in a class.

What will be involved?

Studying   The children can look up information in books such as P. Bailey, *They Can Make Music*, and M. Mandell and R. E. Wood, *Make Your Own Musical Instruments*.

See bibliography: 2, 20.

They might also find out about instruments from other countries, for example the Latin-American morache and tubo, and Far Eastern wood blocks.

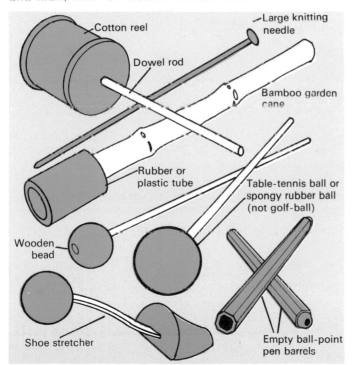

Cotton reel
Large knitting needle
Dowel rod
Bamboo garden cane
Rubber or plastic tube
Table-tennis ball or spongy rubber ball (not golf-ball)
Wooden bead
Shoe stretcher
Empty ball-point pen barrels

Tapping a rhythm with minstrel bones

Some could do historical research and test the so-called 'minstrel' bones, which were beef rib-bones. These cleaned make good clear notes when tapped together. Bones have always been used in music making.

See bibliography: 2.

33

Imagining   They should look about for everyday objects which they observe to have the right properties, for example:

A line of wire coat-hangers
Several simple plastic pencil sharpeners shaken in a jar
A small tin of Biro caps shaken
A pair of walnut shell halves or seashells (such as cockle-shells) drilled and threaded to make castanets or just clapped together

Playing some jingles and tappers

Collecting   Dried peas, beads, yoghurt pots (for shakers), coconut shells, plastic bottles.

Constructing   Sawing, sticking, drilling, decorating the 'instruments'.

See bibliography: 2.

The finale should be a performance, either with melodic instruments or with a record or tape-recording, in which all the jingles and tappers will be played as a percussion section before being put on display.

10 Tuning-fork and tape-recorder

Neither of these is in fact a musical instrument, but both are extremely valuable in the study of science from musical instruments.

The tuning-fork

Carry out the experiments in this section yourself, and then let children make guided discoveries along the same lines.

Scientifically it will help to crystallize much of the work which has been learnt from other apparatus. (See also pages 4-16.)

What does a tuning fork do? It produces sound. Can one find out if this is a result of vibration?

To avoid any damping effect, always hold the fork at the 'waist' with finger and thumb.

Investigations
Getting the fork to sound Do this either by tapping the prong on a hard rubber or a bit of soft scrap wood (not the table), or by pinching the two prongs together and letting go smartly. Then feel the prongs.

Make the fork sound and hold it with one prong against the edge of a saucer. Repeat this with a little water in the saucer.

Make the fork sound and hold it upside down with the prongs just dipping into water in a dish or plastic lunch-box. (Dry the fork after you do this as it is made of steel and may rust.)

Dipping a sounding tuning-fork in water

What results do you observe? What earlier discovery does this reinforce, that is, what hypothesis is this testing? Does it fit?

The order of pitch Choose one tuning-fork from a range of several, if possible from a set of eight (expensive). Draw round it as closely as possible on squared or lined paper. Then get it to sound clearly, and try to memorize the pitch. (By singing it? See page 5.)

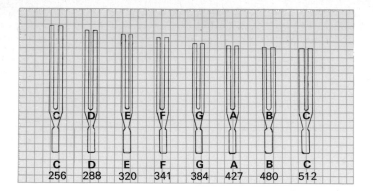

C	D	E	F	G	A	B	C
256	288	320	341	384	427	480	512

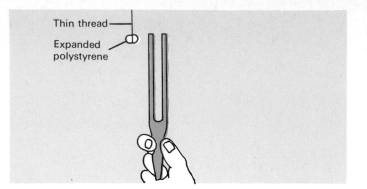

Thin thread —
Expanded polystyrene

Then take another fork, but do not look at the letter on it which indicates its pitch. Measure this fork against the first one, and on this basis predict whether its note will be higher or lower. Then test by sounding it.

You might do this yourself with a complete set, then arrange them in pitch order, and finish by drawing round each. The result could be rather satisfactory. Weigh the forks too and see if there is a connection.

Children will enjoy working with tuning-forks, but it is quite likely that they will have difficulties deciding about the order of pitch.

Can you test with pupils to see if this ability is acquired through training, or is a function of the developmental stage, or a personal characteristic— some can, some can't? This is important to know, so that you can allow for it in other music work.

In fact, much of the science here can be grasped without exact pitch discernment.

Vibration Test the vibration effect by holding a sounding tuning-fork so that the side of one prong touches a very small piece of expanded polystyrene hanging from a fine thread.

The sound-box effect Investigate the effect of a sound-box by holding a vibrating tuning-fork with the end of its stem alternately held down on the table and up in the air. Repeat this with pieces of wood, hollow

boxes, tins, a coffee jar with the lid on. Try carefully holding the end of the tuning-fork against a guitar sound-box.

How do you explain the differences to yourself? How do children explain them? See how they think about this.

The tape-recorder

A reel-to-reel tape-recorder can help with questions of pitch and frequency, that is the rate of vibration.

Record a single note (say middle C on the piano), played several times, at the three speeds of a tape-recorder, one after the other. Then play all three recordings at one speed. Do this first to yourself, and later to children. One piece will sound right. The others will show exactly what difference frequency, or the number of vibrations per second, makes in sounds.

Ask a child to play a small tune, say on a mouth-organ, three times over. Record it at a different speed each time. Play it back at any one speed. Let the children guess whether the recording was made at the correct speed, or slow, or fast.

The recordings at different speeds suggested here would have to be made on a reel-to-reel tape-recorder, but the results needed for class work could easily be re-recorded with a cassette.

11 Making a one-stringed guitar

This is a considerable enterprise in its own right. Here are detailed instructions for this particular instrument, so that you or a colleague can produce a classroom guitar, with some assistance from pupils at any stage you find suitable.

This guitar is similar to the one shown in Science 5/13 *Science from toys*, page 15. It is most suitable for children aged between nine and thirteen.

See bibliography: 14.

Materials and tools

3-ft length of $1\frac{1}{2} \times \frac{5}{8}$-in whitewood	Drawing pins
$\frac{1}{4}$-in diameter dowel rod	Matchsticks
$\frac{1}{2}$-in diameter dowel rod	One $2\frac{1}{2}$-in nail
Cardboard box with thin	$\frac{1}{2}$-in screws
rigid sides	Sellotape
Guitar string	Glue
$\frac{3}{4}$-in hardboard pins (panel pins)	Small tenon saw
Hand drill or brace	Bradawl
High-speed drill bits ($\frac{1}{8}$-in and $\frac{11}{32}$-in)	Craft knife
Small hammer	Scissors
Medium-sized screwdriver	Junior hacksaw

Organization

Adapting the classroom Few classrooms have work benches. You may therefore need to consider how to adapt and improvise from the furniture and equipment available. There are two main problems:

Avoiding damage to classroom fittings and furniture.

The absence of suitable equipment. How, for example, can you drill holes in dowel rod without a vice?

Examine the classroom and consider its suitability for craft work, and how best it can be adapted for this activity.

Safety This activity demands that the teacher must make sure that the children use the tools safely. Try to anticipate the main dangers involved and ways of overcoming them.

For example, when using a craft knife the child should direct the cutting movement away from himself and any other children nearby. When sawing, the wood may slip, resulting in nasty cuts. It can be secured by using a G-cramp or a bench-hook. See also Science 5/13 *Working with wood* (page 15), *Science from toys* (page 49), and *Science from wood* in this series.

What other dangers can you envisage and how can they be tackled?

See bibliography: 6, 14, 17.

Determining the position of the frets

If a particular note is required from the open string, use a piano, tuning-fork or pitch pipe. The fret positions can be obtained by pressing the tuned string at various points until the required note is produced.

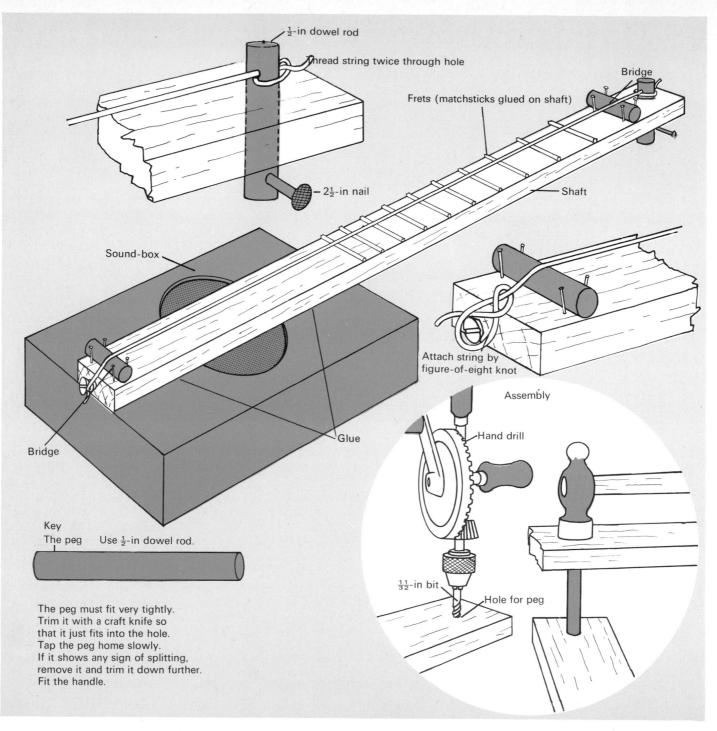

½-in dowel rod

Thread string twice through hole

Bridge

Frets (matchsticks glued on shaft)

2½-in nail

Shaft

Sound-box

Attach string by figure-of-eight knot

Bridge

Glue

Assembly

Hand drill

$1\frac{1}{32}$-in bit

Hole for peg

Key

The peg   Use ½-in dowel rod.

The peg must fit very tightly.
Trim it with a craft knife so
that it just fits into the hole.
Tap the peg home slowly.
If it shows any sign of splitting,
remove it and trim it down further.
Fit the handle.

While positioning the frets you and the children may notice relationships between the lengths of the vibrating sections of the strings. For example, what is the relationship between the vibrating length of the open string and the length of string which produces the *same* note one octave higher? What other relationships of this kind can you find?

How far do you agree that children should not make the guitar unless they understand the principles underlying playing it? Have you discovered that children put any general principles into practice when constructing this instrument?

Testing alternatives

You could try:

Changing the string; try wire, rubber, elastic.

Using the shaft without a sound-box.

Changing the sound-box: what about a biscuit tin or cigar box?

Using a sound-box without any hole.

Making openings in the sound-box of different sizes and in different positions. Cardboard sound-boxes are best for this because cardboard can easily be removed and replaced.

Such activities are scientifically fruitful. They are bound to result in new observations, which in turn may provoke further experimentation.

Tests of the different alternatives should be as fair or 'controlled' as possible. New modifications should be tested one at a time, for it is difficult to establish the effects of any one when several are tried out simultaneously. Tests of alternatives should be carried out under similar conditions. For example, would a test to determine the effect of two different sound-boxes produce useful results if they were resting on different surfaces?

Did your observations of the effect of one sound-box enable you to predict the effect of another?

There are limits to the information you can obtain from such broadly based tests. Consider the following:

Perry, aged eleven, was dissatisfied with the effect produced by a sound-box made from a cardboard container. He replaced it with a square biscuit tin without a lid, which he screwed into the shaft through two holes in its base. This resulted in a louder sound of a different quality. The explanation he gave was: 'One box was made of tin and the other of cardboard.'

What other possibilities might you suggest to Perry? How, if at all, could you put these to the test?

**Possible further study
extensions**

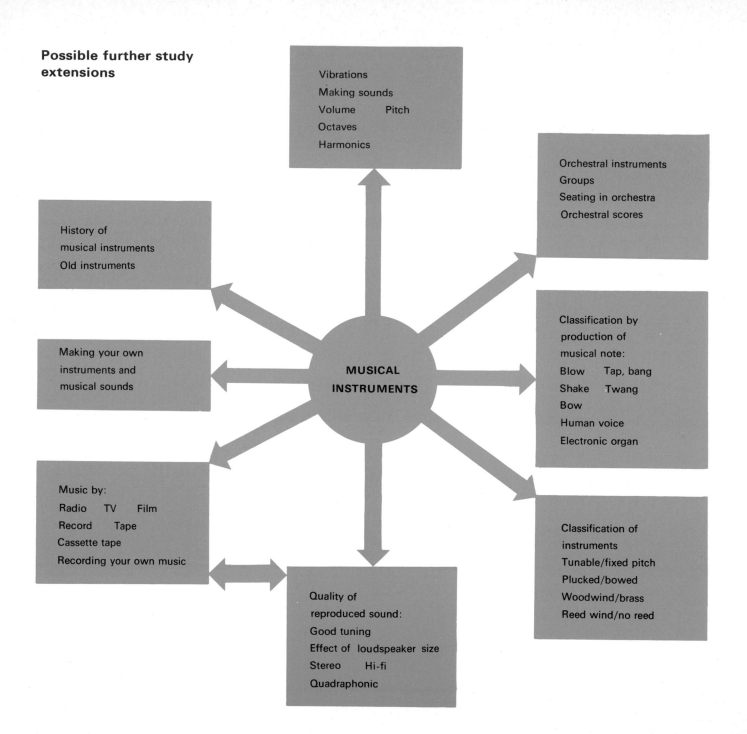

Vibrations
Making sounds
Volume   Pitch
Octaves
Harmonics

Orchestral instruments
Groups
Seating in orchestra
Orchestral scores

History of
musical instruments
Old instruments

Classification by
production of
musical note:
Blow   Tap, bang
Shake   Twang
Bow
Human voice
Electronic organ

Making your own
instruments and
musical sounds

**MUSICAL
INSTRUMENTS**

Music by:
Radio   TV   Film
Record   Tape
Cassette tape
Recording your own music

Classification of
instruments
Tunable/fixed pitch
Plucked/bowed
Woodwind/brass
Reed wind/no reed

Quality of
reproduced sound:
Good tuning
Effect of loudspeaker size
Stereo   Hi-fi
Quadraphonic

Bibliography

For children to use

1 Brace, G. (1968) *The Story of Music*. Ladybird Books. Wills & Hepworth.
2 Mandell, M. E. and Wood, R. E. (1970) *Make Your Own Musical Instruments*. Bailey Bros. This book contains a fantastic collection of ideas for making simple shakers, tappers, drums, bongos, chimes, etc. Very practical and decorative. See especially page 59.
3 Rees, A. (1966) *Musical Instruments*. Ladybird Books. Wills & Hepworth. A most useful book, even for adults.
4 Sanday, A. P. (1975) *Sounds*. Ladybird Books. Wills & Hepworth.

For direct work with children

5 Brett, B. and Ingman, N. (1972) *The Story of Music*. Ward Lock. Includes The Who, Jimi Hendrix, Creedence Clearwater and the Moog synthesizer; some good material; more pretty than useful.
6 Diamond, D. (1976) *Science from wood*. Nuffield and Chelsea College Teaching Primary Science series. Macdonald Educational.
7 Headington, C. (1965) *The Orchestra and its Instruments*. Bodley Head. Written for secondary school pupils but not difficult.
8 Schools Council Science 5/13 (1973) *Change*, Stage 3. Macdonald Educational. See pages 27, 28.
9 Schools Council Science 5/13 (1972) *Early experiences*. Macdonald Educational. See pages 31–39.
10 Schools Council Science 5/13 (1973) *Holes, gaps and cavities*, Stages 1 and 2. Macdonald Educational. See pages 19–21.

11 Schools Council Science 5/13 (1973) *Like and unlike*, Stages 1, 2 and 3. Macdonald Educational. See pages 6, 43, 44.
12 Schools Council Science 5/13 (1973) *Metals*, Stages 1 and 2. Macdonald Educational. See pages 16, 19. Gives lengths for chime bells.
13 Schools Council Science 5/13 (1973) *Ourselves*, Stages 1 and 2. Macdonald Educational. See pages 22–25.
14 Schools Council Science 5/13 (1972) *Science from toys*, Stages 1 and 2 and background. Macdonald Educational. See pages 11–15.
15 Schools Council Science 5/13 (1974) *Using the environment 1 Early explorations*. Macdonald Educational. See page 25.
16 Schools Council Science 5/13 (1974) *Using the environment 2 Investigations Part 2*. Macdonald Educational. See page 99.
17 Schools Council Science 5/13 (1972) *Working with wood*, Stages 1 and 2. Macdonald Educational. See pages 16, 17. Scanty.

For further information and ideas

18 Allen, G., Brown, V. W., Southam, H. and Tuke, E. M. (1965) *Scientific Interests in the Primary School*. National Froebel Foundation. See pages 22, 23. List of materials and instruments.
19 Ashhurst, W. (1971) *A New Introduction to Physics*. John Murray. See pages 268–269.
20 Bailey, P. (1973) *They Can Make Music*. Oxford University Press. Instruments and teaching methods for handicapped children; also a good bibliography.
21 Cameron, K. and Mathieson, M. (1973) *We Make Music, A Series of Films*. Educational Foundation for Visual Aids.

22 Bainbridge, J. W., Stockdale, R. W., and Wastnedge, E. R. (1970) *Junior Science Source Book*. Collins. See pages 240, 246.

23 Brown, M. and Precious, G. W. (1968) *The Integrated Day in the Primary School*. Ward Lock Educational. See pages 47, 64, 77, 98–100. Good list of instruments.

24 Garnett, H. (1971) *Practical Music Making with Juniors*. Schoolmaster Publishing Co Ltd.

25 Hosier, J. (1961) *Instruments of the Orchestra*. Oxford University Press.

26 Ivimey, G. (1974) *Playing, Working, Growing*. Temple Smith. See pages 116, 121, 122.

27 Jacobs, A. (1970) *New Dictionary of Music*. Penguin.

28 James, A. (1964) *Simple Science Experiments*. Schofield & Sims.

29 Jardine, J. (1964) *Physics is Fun*, Book 2. Heinemann Educational. See pages 64-87. Meant for Scottish thirteen and fourteen year olds, this is a delightful book, with many practical ideas and clear explanations.

30 Mitchell, M. and Youngs, M. (1965) *The Roots of Science*. Blackwell. See pages 34, 72, 86.

31 Nuffield Junior Science (1967) *Apparatus*. Collins. See pages 204–219.

32 Nuffield Junior Science (1967) *Teachers' Guide 1*. Collins. See pages 118–124, 163, 254.

33 Nuffield Junior Science (1967) *Teachers' Guide 2*. Collins. See pages 17–18, 29–30, 62–65, 135–140.

34 Rainbow, B. (ed.) (1964) *Handbook for Music Teachers*, Book 1. Novello. See pages 116 ff.

Contains a section by Kathleen Blocksidge on the making of simple musical instruments.

35 Roberts, R. (1965) *Musical Instruments Made to be Played*. Dryad Press.

36 Tufts, N. P. (1962) *The Art of Handbell Ringing*. Jenkins. Contains some useful information about bells and suggestions for junior handbell ringing groups.

37 Vries, L. de (1958) *The Book of Experiments*. John Murray. See pages 48–60, especially: nail piano, water trombone, page 55; milk bottle organ, page 56; wine glass, page 58; cardboard-tube xylophone, page 59.

38 Vries, L. de (1963) *The Second Book of Experiments*. John Murray. See pages 25–49, especially: xylophone, page 32; Swanee whistle, page 34; tea-chest bass, pages 35–36.

Equipment catalogues

39 E. J. Arnold: Classroom and Environmental Science 1976. Arnold Science Unit: Sound and Music (with activity sheet master set and teacher's notes).

40 Arnold Music Project Kit (with instructions for children aged between eleven and thirteen). Separate items for study of music: tuning-forks, handbells, chime bar, triangle, string bars, etc.

41 Stewart Plastics Limited, Purley Way, Croydon, CR9 4HS. Household ware catalogue: excellent plastic containers, eg sandwich boxes, screwtop jars, beakers.

Index

Acknowledgements
The authors and publishers gratefully acknowledge the
generous help given by:

The staff and pupils of:

Little Ealing Middle School,
London Borough of Ealing, London W5
The students of Thomas Huxley College, London W3

Illustration credits
Photographs
Crown Copyright Victoria and Albert Museum, page 20
Terry Williams, all other photographs

Line drawings by GWA Design Consultants

Cover design by GWA Design Consultants